Things to Count

Animals

6 – 10

Illustrated by Gary Mohrman

Teaching & Learning Company

1204 Buchanan St., P.O. Box 10
Carthage, IL 62321-0010

Table of Contents

Six 5

Find six 6

Find six 7

Write 8

Match six 9

Seven 10

Find seven 11

Find seven 12

Write 13

Match seven 14

Eight 15

Find eight 16

Find eight 17

Write 18

Match eight 19

Nine 20

Find nine 21

Find nine 22

Write 23

Match nine 24

Ten 25

Find ten 26

Find ten 27

Write 28

Match ten 29

Match picture to word 30

Match picture to numeral 31

Match 32

Cover art by Gary Mohrman

ISBN No. 1-57310-224-5

Printing No. 987654321

Teaching & Learning Company
1204 Buchanan St., P.O. Box 10
Carthage, IL 62321-0010

How to Use This Book

The activities in this book have been created to give children a variety of experiences in working with numbers. The first page introduces the number and is followed by activities which stress counting, matching and writing. The activities can be used to supplement your instructional program, for review, extra practice, take-home pages or for fun fillers. The material is flexible enough to use in a variety of ways and, of course, you may wish to create custom pages suited to a child's individual need based upon any one of the ideas in this book.

As the directions for each exercise in this book are simple and easy-to-follow, these materials are well-suited for use in a setting where slightly older children might be available to help the youngest students. Children who can read the directions themselves, or to whom the material has been explained, should be able to complete the work independently.

Booklets/Number Line

The pages which introduce each number can be used to create a classroom display or take-home. Copy the pages and tape together to create a number line. Children can color each page to create a valuable teaching aid for your room. You could also make a take-home booklet or number line for each child.

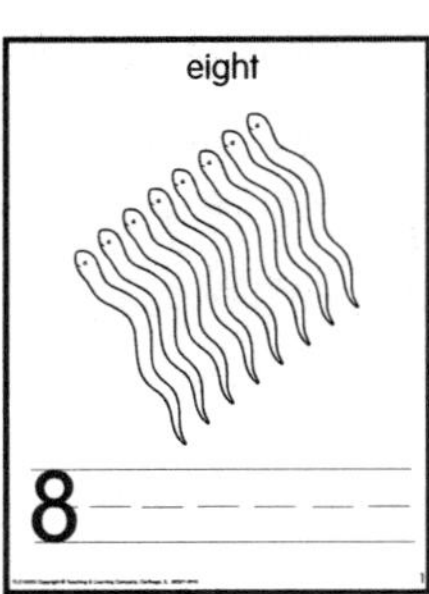
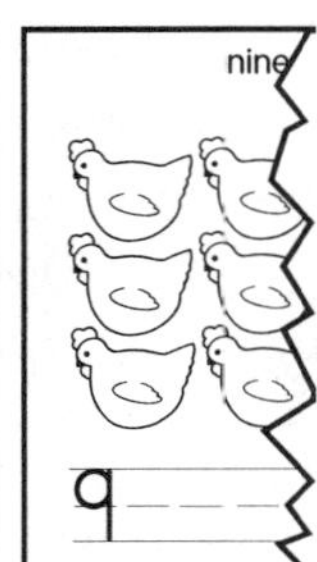

The artwork in this book is deliberately simple, bold and easy for young children to color. These same attributes make it ideal for copying to cut out for bulletin board or wall displays, booklet or folder covers, mobiles, name tags, overhead projections, etc. For stand-up figures, mount the artwork to tagboard, color and cut out. Make a simple V-shaped base with slots and insert the figure.

Stand-Up Figures

These sturdy stand-up figures can be used in a variety of counting, sorting, matching and grouping activities.

six

6

Find six .

Find six 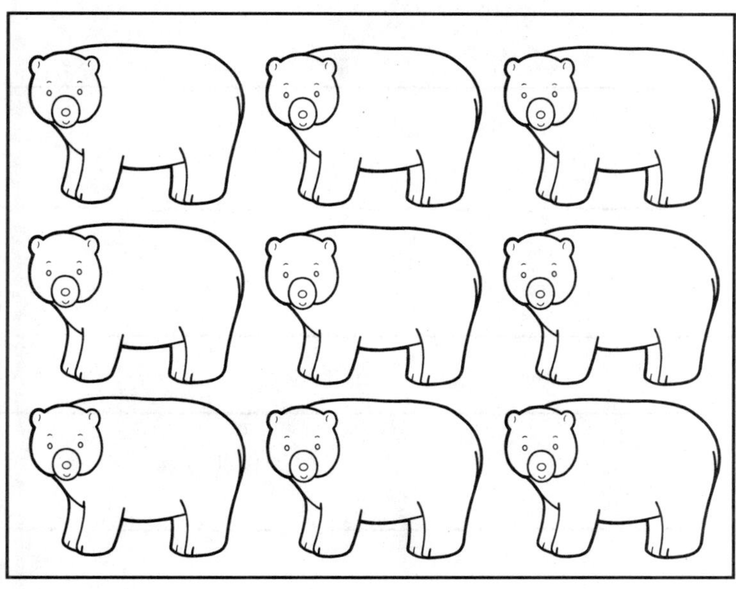

Write.

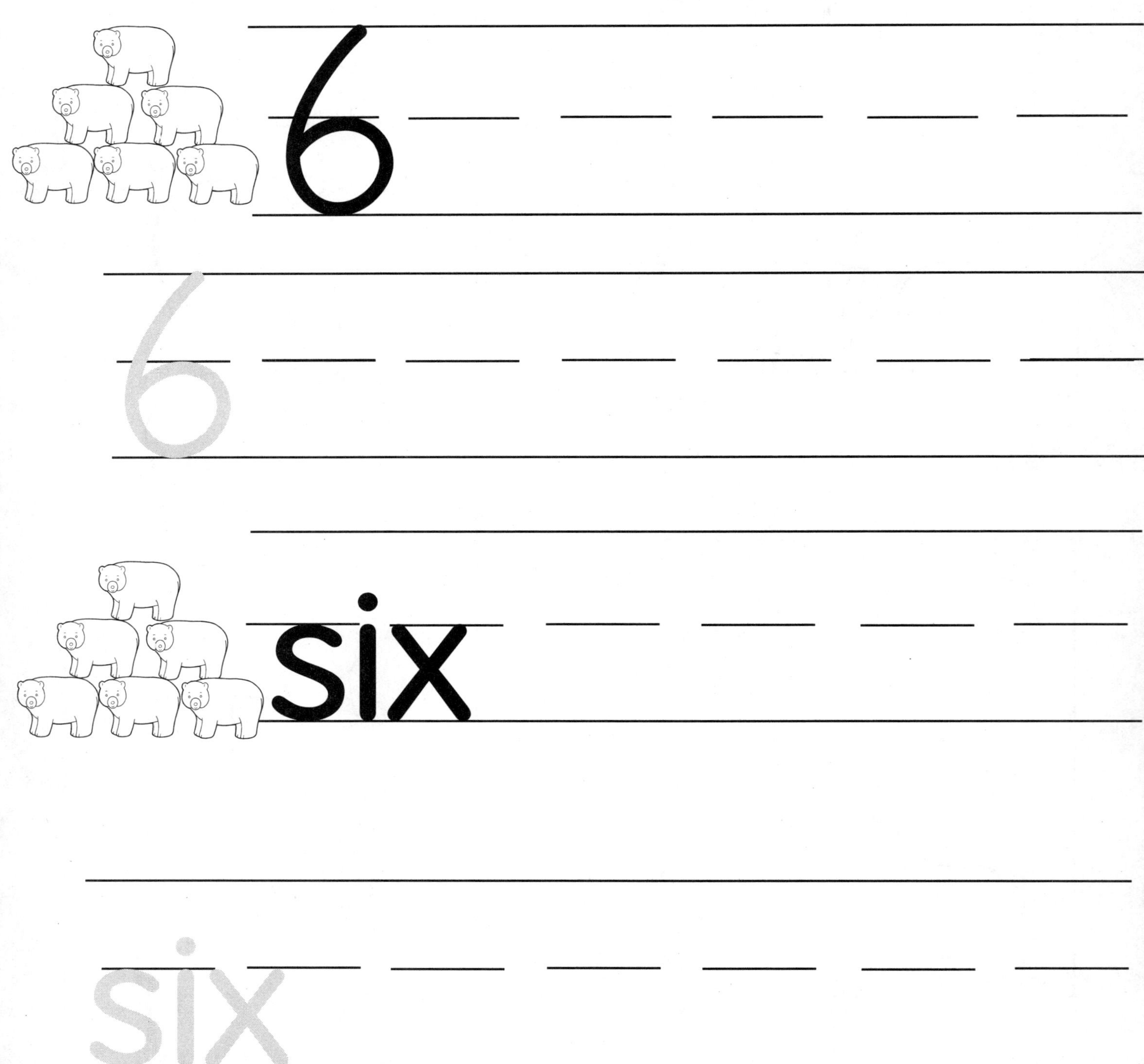

Match  six.

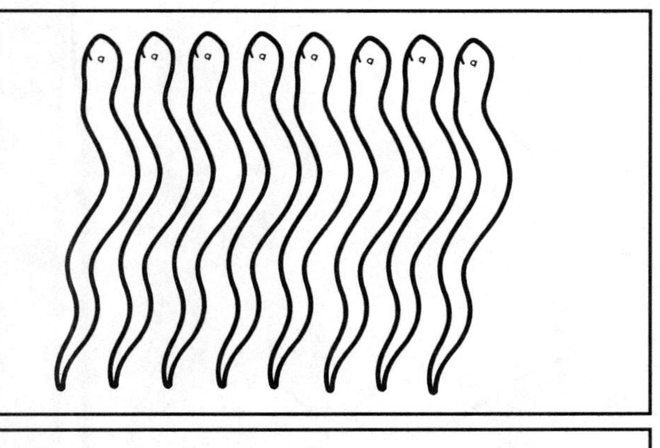

stick

sick

sticks

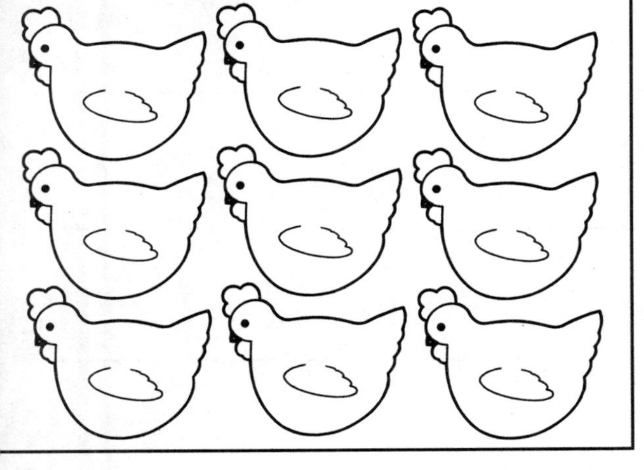

six

9

seven

7 - - - - - - - -

Find seven

Find seven 

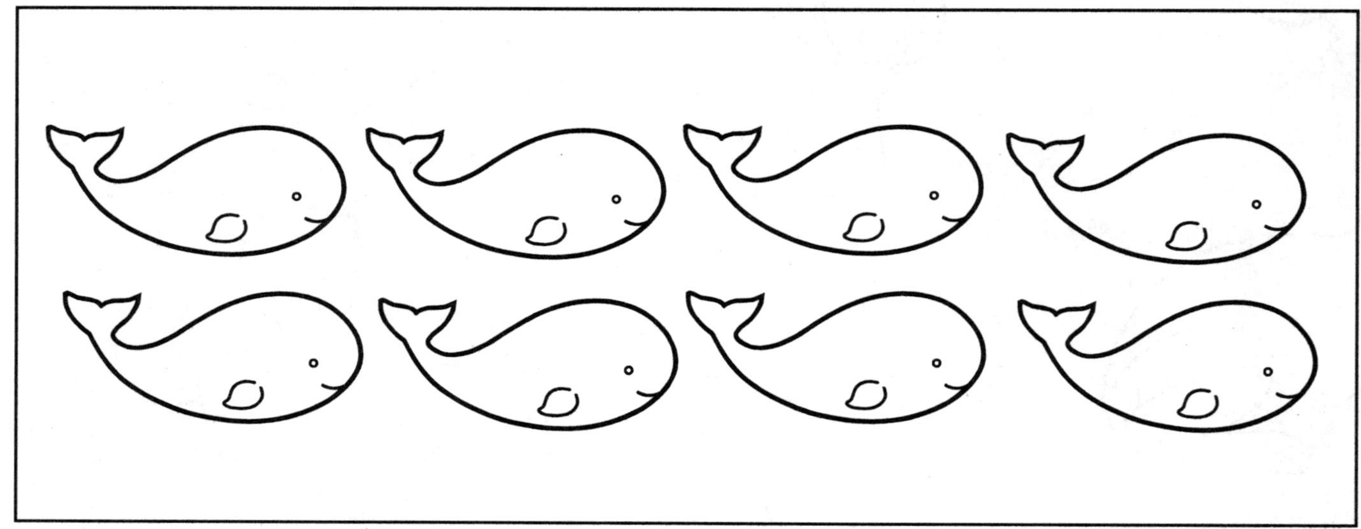

Write.

Match 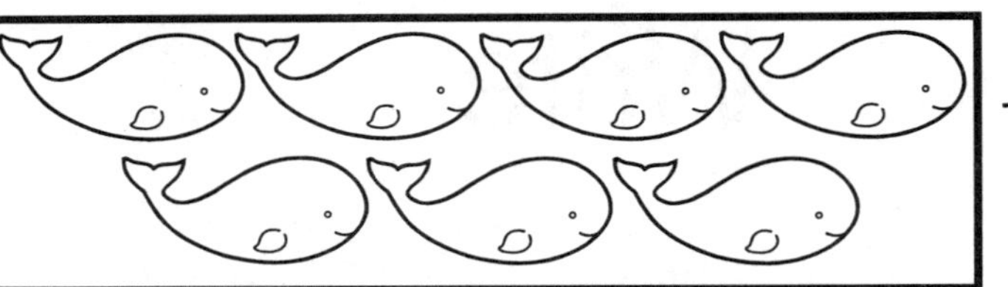 — seven .

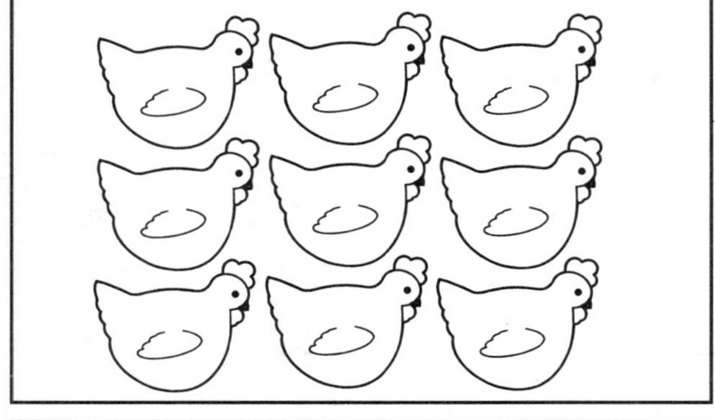

even

seven

never

ever

eight

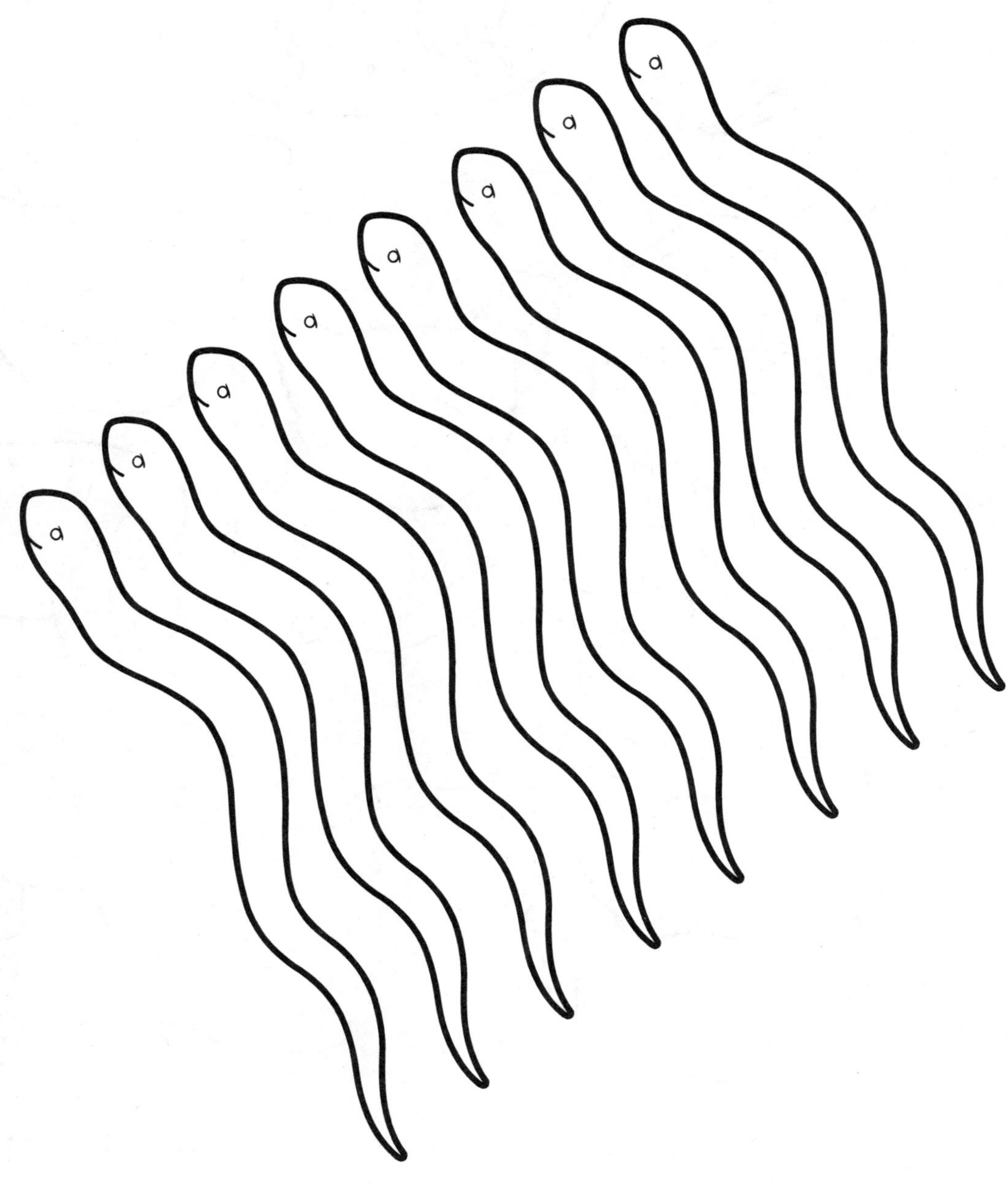

8

Find eight

Find eight 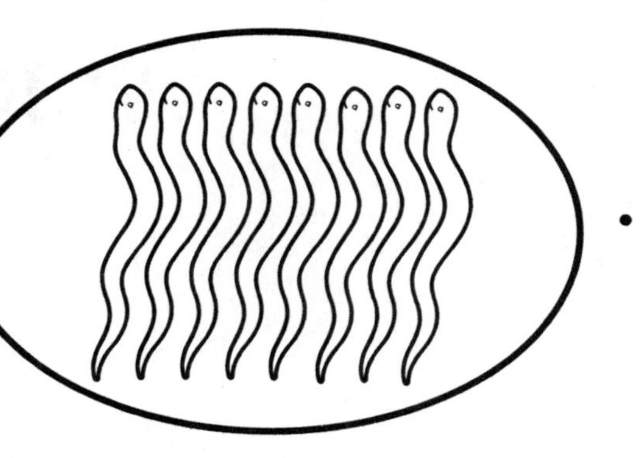

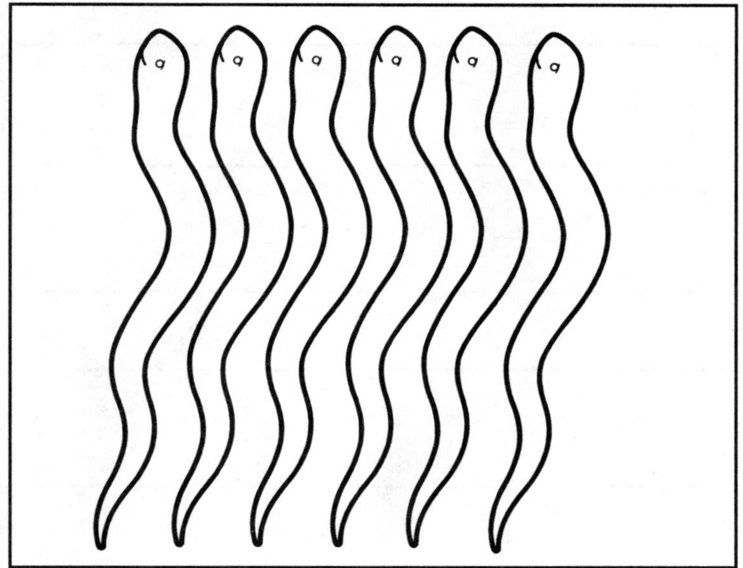

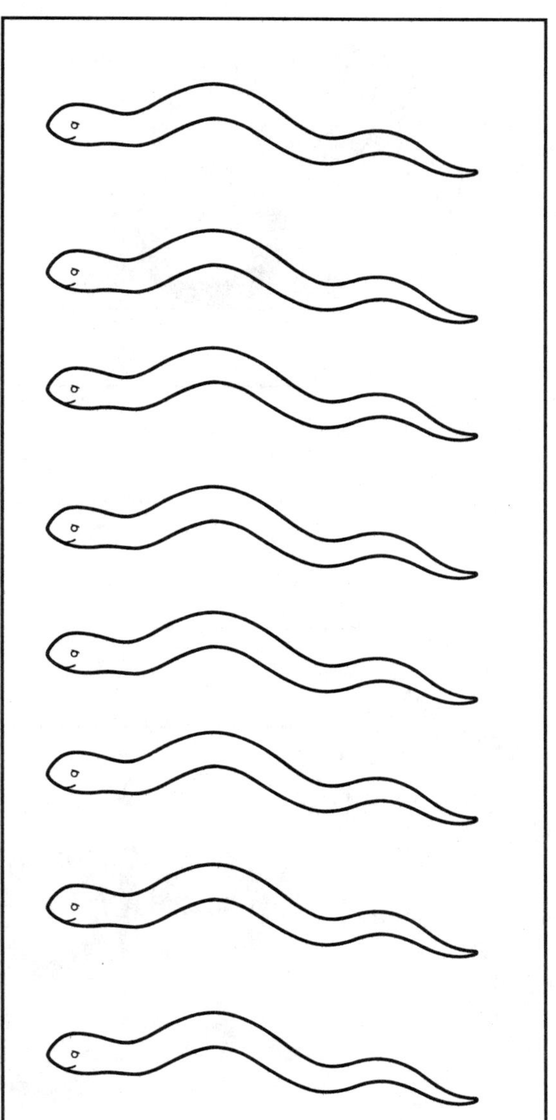

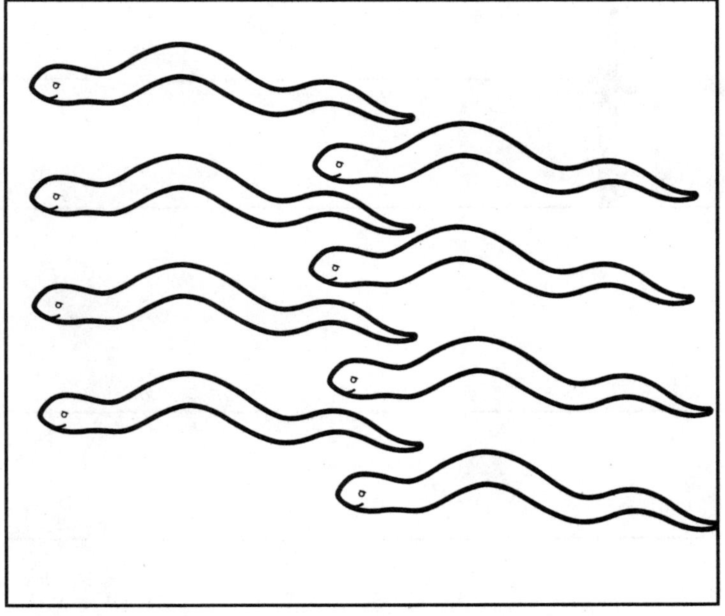

Write.

8

8

eight

eight

Match 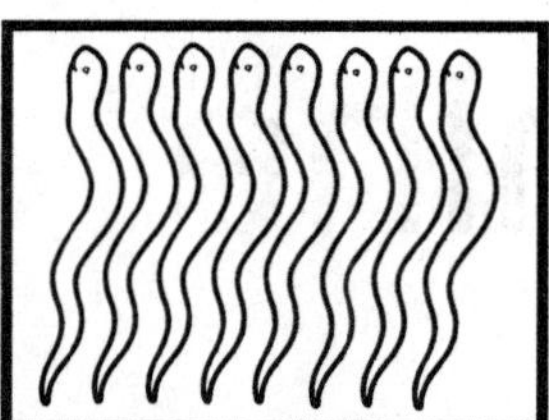 — eight .

 sight

egg

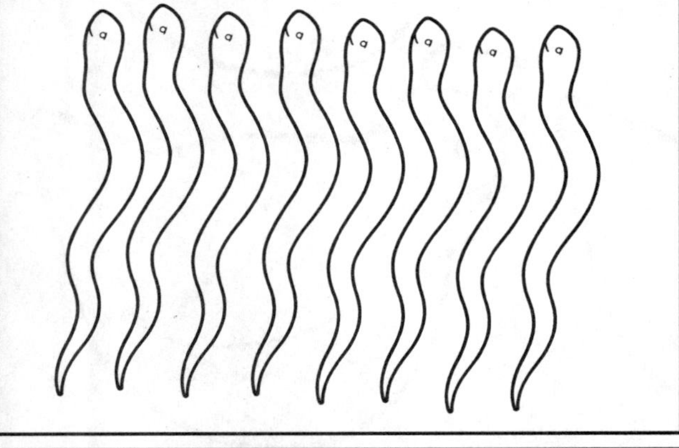

eine

eight

nine

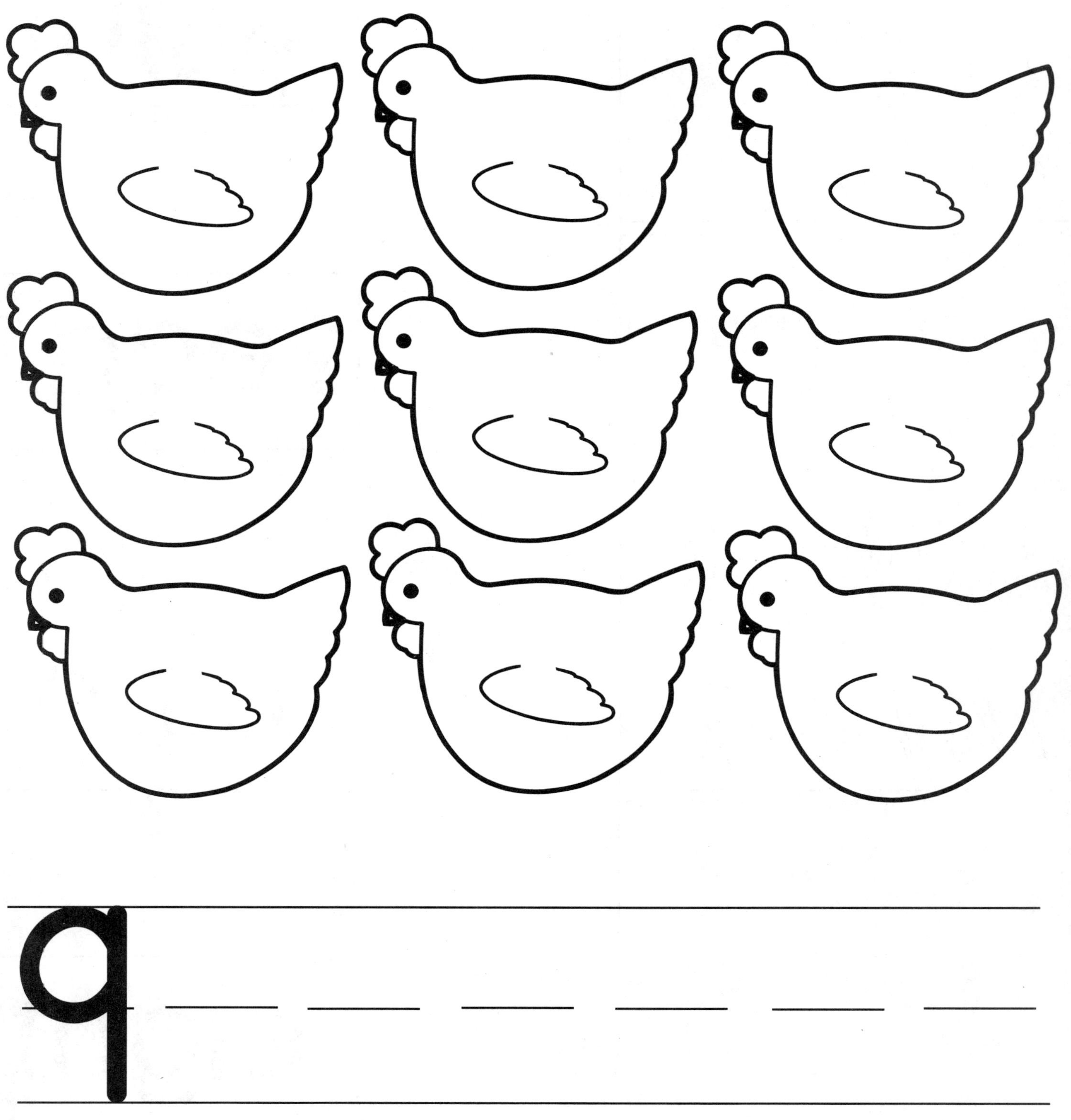

q

Find nine 

Find nine 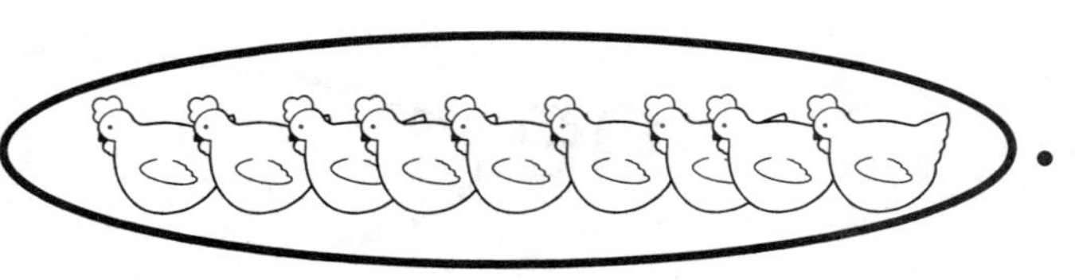.

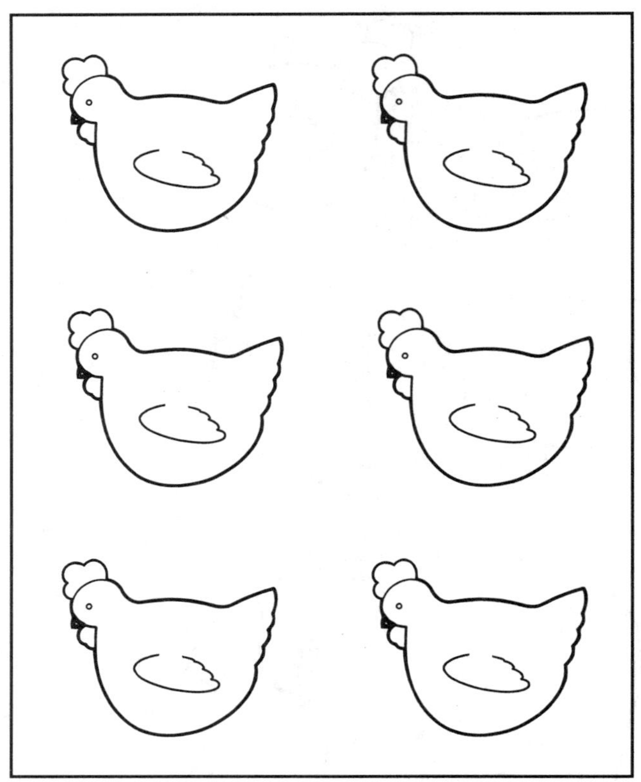

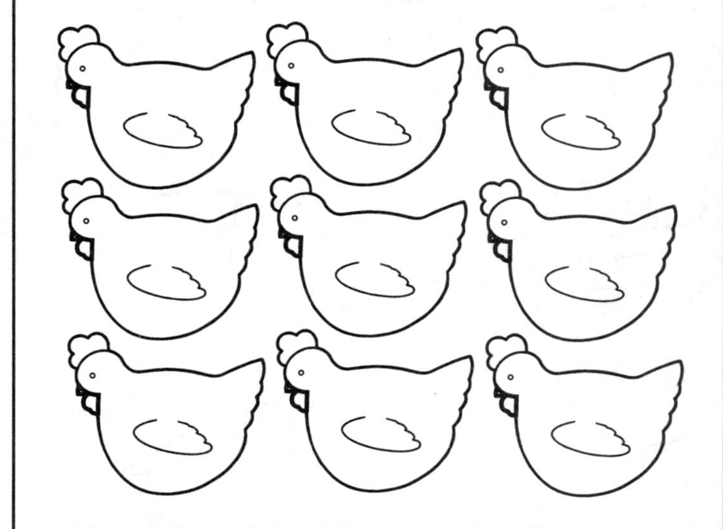

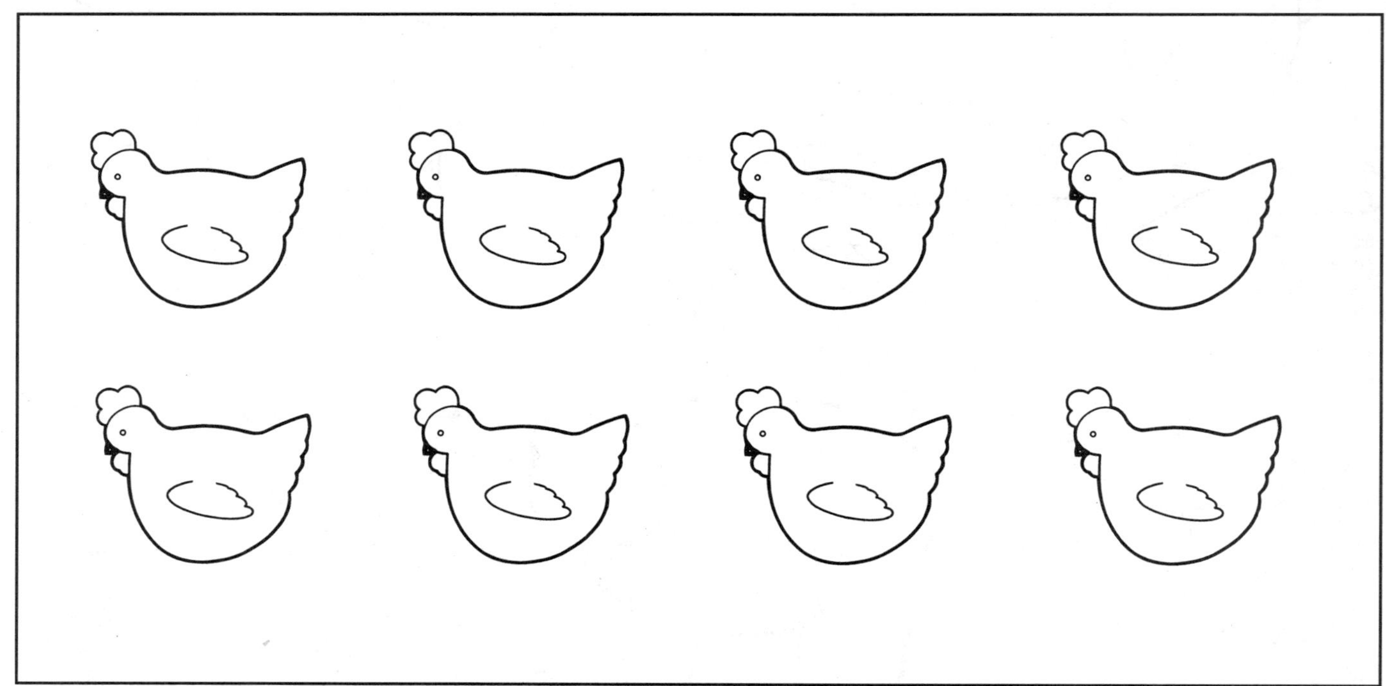

Write.

Match  — nine .

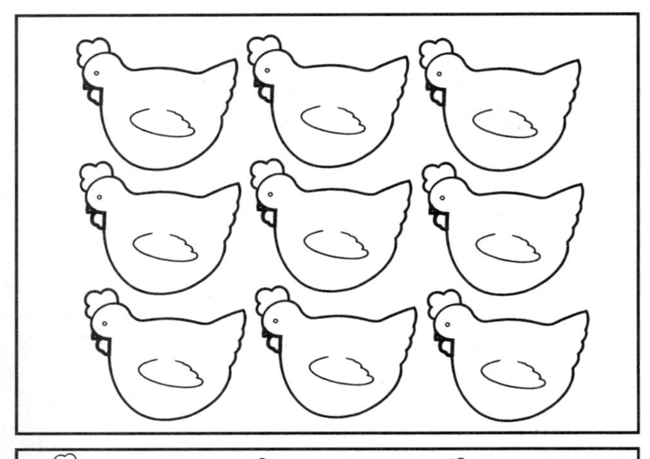

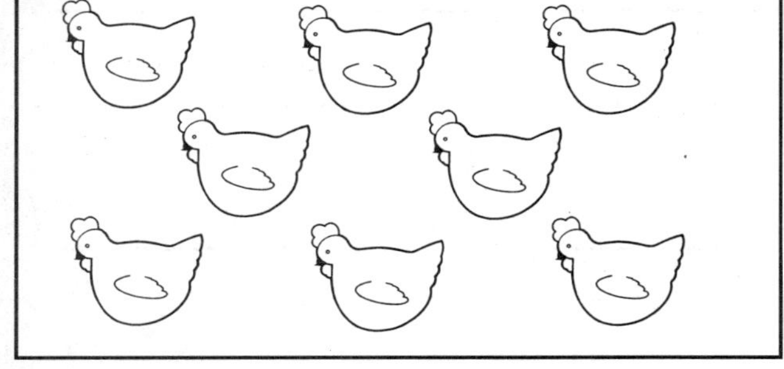

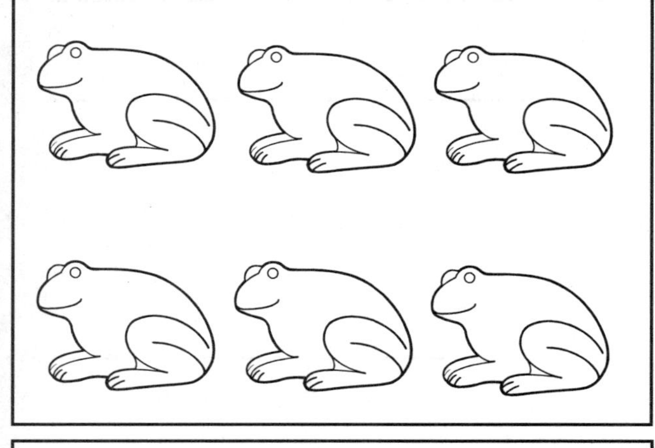

line

fine

nine

nick

ten

10 — — — — —

Find ten

Find ten

Write.

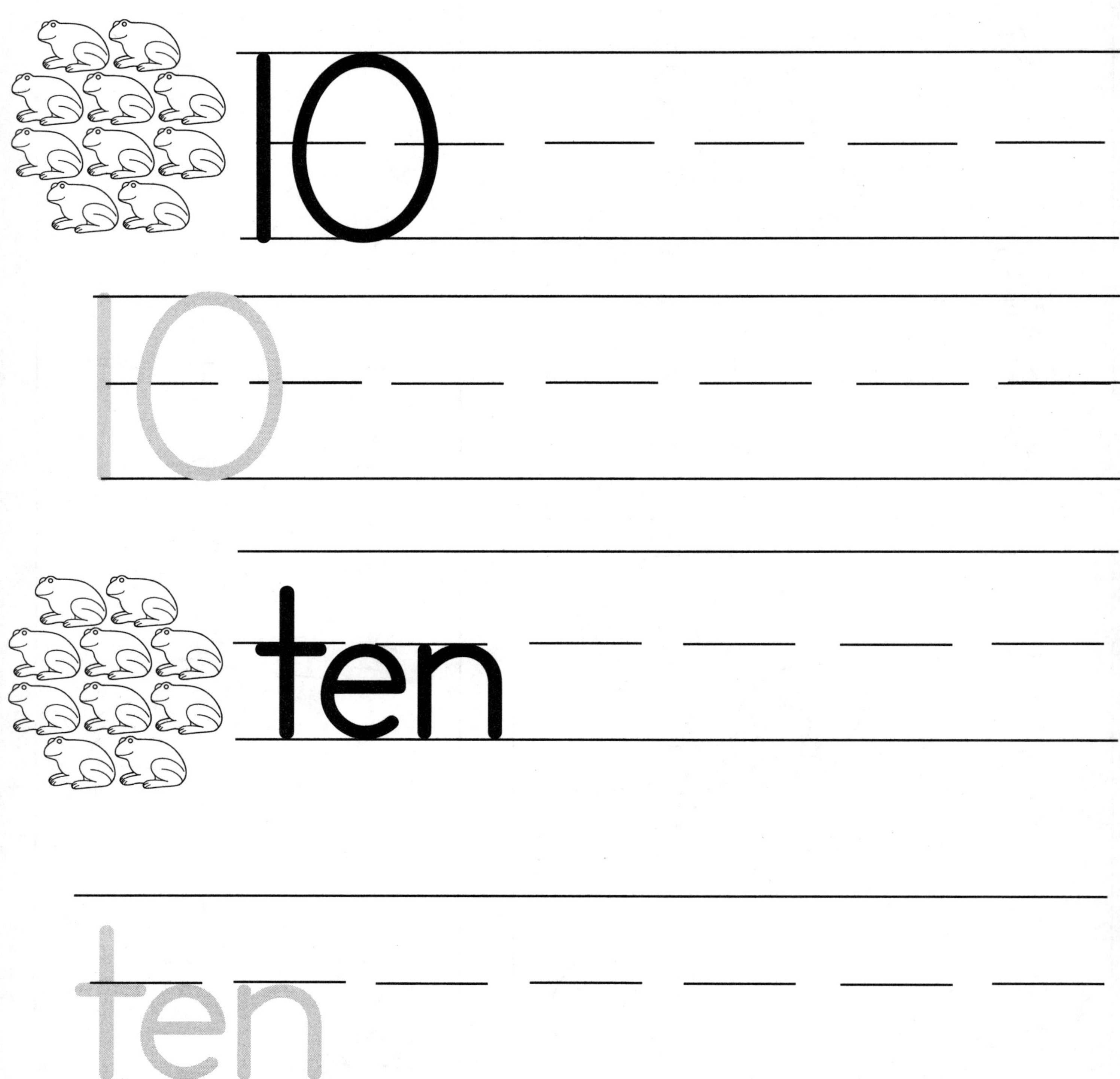

Match  — ten .

net

ten

tex

two

Match 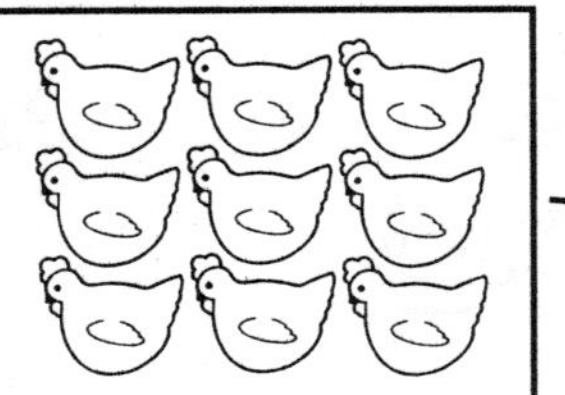 — nine .

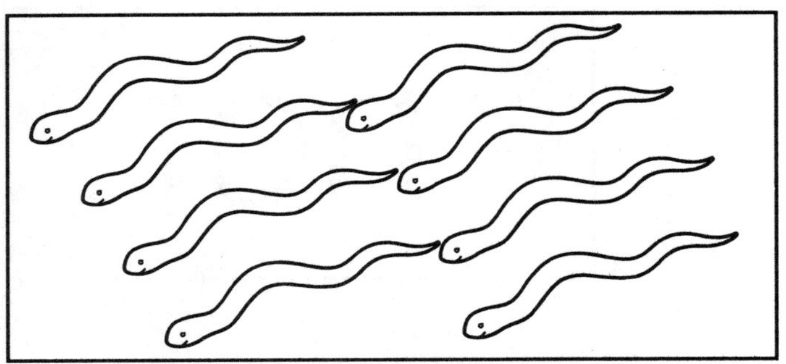

six

nine

eight

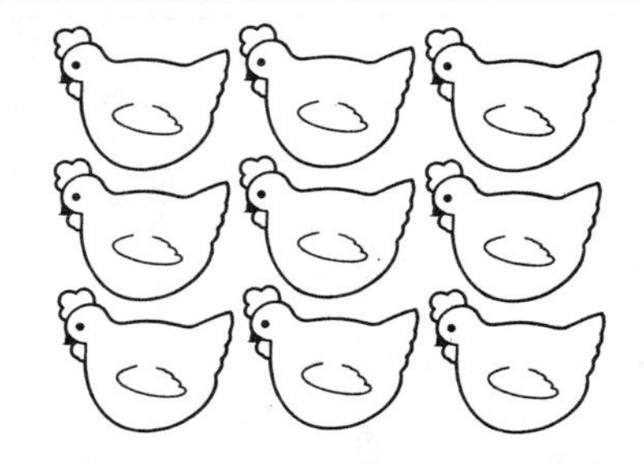

ten

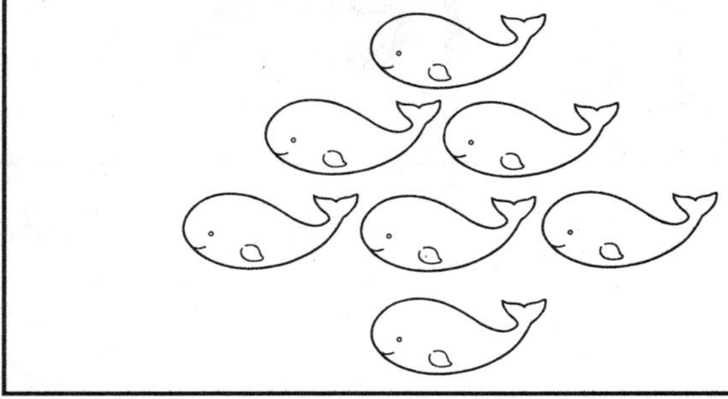

seven

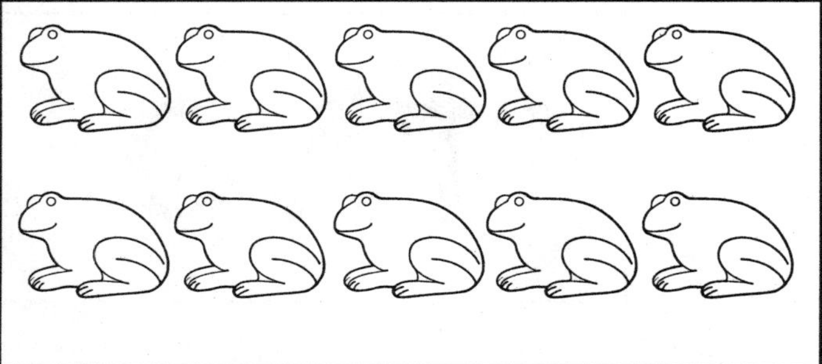

Match  — 6 .

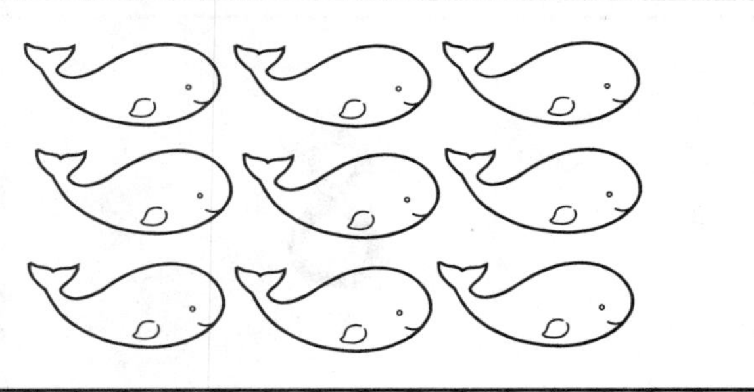

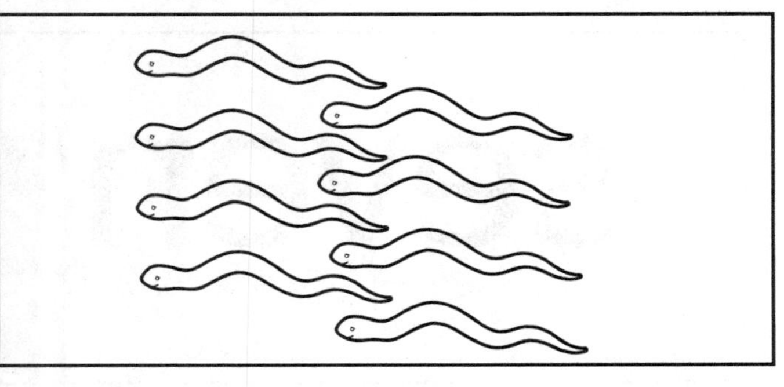

7

6

q

10

8

Match.

7

four

4

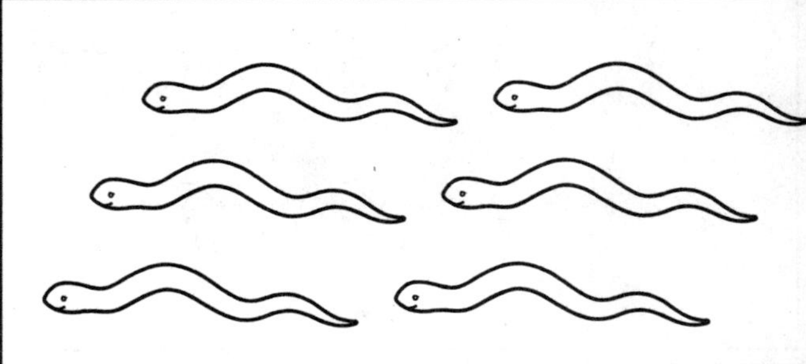

8

10

six

seven